Food Innovation and Engineering (FOODIE) Asia Conference 2019

Singapore
5 – 6 December 2019

ISBN: 978-1-7138-0569-4

Printed from e-media with permission by:

Curran Associates, Inc.
57 Morehouse Lane
Red Hook, NY 12571

Some format issues inherent in the e-media version may also appear in this print version.

Printed with permission by Curran Associates, Inc. (2020)

AIChE
120 Wall Street, FL 23
New York, NY 10005-4020

Phone: (800) 242-4363
Fax: (203) 775-5177

www.aiche.org

Additional copies of this publication are available from:

Curran Associates, Inc.
57 Morehouse Lane
Red Hook, NY 12571 USA
Phone: 845-758-0400
Fax: 845-758-2633
Email: curran@proceedings.com
Web: www.proceedings.com

TABLE OF CONTENTS

SESSION 4: HEALTH AND THE GUT MICROBIOME

SESSION 5: FUTURE FOODS, INNOVATIONS AND NOVEL NUTRITIONAL SOLUTIONS

POSTER SESSION

Planetary Talk: Healthy Diet from Sustainable Food System

FOODIE Asia
FOOD INNOVATION AND ENGINEERING CONFERENCE

Conference: Food Innovation and Engineering Asia Conference

Year: 2019

Proceeding: 2019 Food Innovation and Engineering (FOODIE) Asia Conference

Group: General Submissions

Session: Plenary Session I

Time: Thursday, December 5, 2019 - 9:15am-9:40am

Authors:

Agustina, R., Faculty of Medicine Universitas Indonesia

Healthy food is a key to optimize human health and achieve environmental sustainability on Earth. However, current transition and practices in food production and consumption are threatening both people and planet. Therefore, taking â€œearth system approachâ€ in defining a sustainable food system is crucial. The EAT-Lancet Commission on Healthy Diet and Sustainable Food System recently published a report in 2019 that defined a planetary healthy diet, as a global reference diet for adults, that is healthy for humans and avoid environment degradation to the planet. Shifting into healthier diet based on this reference is likely to result in substantial health benefit and predicted to reduce rates of malnutrition, non-communicable diseases, and about 20% of annual premature death among adults. A set of actions was proposed by the Commission to meet the scientific targets for healthy diets and sustainable food production that allow to stay within the safe operating space for food systems. This action requires a combination of substantial shifts toward great variety of plant-based dietary intake, significant reductions in food losses and waste, and major improvements in food production practices. The Commission proposed five potential strategies for food system transformation. More information on this report can be found in Willet et al. Food in Anthropocene. The Lancet, 2019. Feb 9;393(10171):530.

Invited Talk: Overview of Aquaculture Innovation Centre (AIC)

FOODIE Asia
FOOD INNOVATION AND ENGINEERING CONFERENCE

Conference: Food Innovation and Engineering Asia Conference

Year: 2019

Proceeding: 2019 Food Innovation and Engineering (FOODIE) Asia Conference

Group: General Submissions

Session: Session 1: Technologies for Primary Production

Time: Thursday, December 5, 2019 - 10:45am-11:10am

Authors:

Wee, L. C., Aquaculture Innovation Centre

Aquaculture Innovation Centre â€“ Your Strategic Partner and Service Provider

Countries across the world are battling the worsening impacts of climate change, water scarcity and population growth to find better ways to feed their people. Singapore imports more than 90% of its food. As a result, local food production is becoming a core component of the food security roadmap.

This year the Singapore Food Agency (SFA) has set the target of producing 30% of Singaporeâ €™s nutritional needs by 2030, up from less than 10% today. A move targeted at improving food security and strengthen resilience of Singaporeâ€™s food supply. To get to the â€œ30 by 30 visionâ€, will require our agri-food industry to adopt new solutions to raise productivity, strengthening climate resilience, and overcoming our resource constraints through technologies.

Aquaculture Innovation Centre (AIC) (founded in June 2019) is a consortium of institutions and agencies adopting an inclusive and integrated approach to fulfil the objectives of serving the needs of the local aquaculture industry through collaboration and co-operation, co-sharing of expertise and facilities. The members of AIC include Ngee Ann Polytechnic, Republic Polytechnic, Temasek Polytechnic, James Cook University Singapore, Nanyang Technological University, National University of Singapore, SFA & A*STAR. The Centre serves as an excellent platform to drive aquaculture industry development, facilitate market opportunities and support government policies and regulations. The speaker will be sharing some of the industry engagement statistics and insights.

Invited Talk: Integrated Agrotechnologies and Microbiome Applications in Urban Farming

FOODIE Asia

FOOD INNOVATION AND ENGINEERING CONFERENCE

Conference: Food Innovation and Engineering Asia Conference

Year: 2019

Proceeding: 2019 Food Innovation and Engineering (FOODIE) Asia Conference

Group: General Submissions

Session: Session 1: Technologies for Primary Production

Time: Thursday, December 5, 2019 - 11:35am-12:00pm

Authors:

Swarup, S., National University of Singapore

Food production and food security rely heavily on high levels of crop productivity and resilience of crop production under diverse environmental stresses. In order to better understand and apply the knowledge of plant and their associated microbiomes, we have been using a combination of systems biology approaches to integrate modern life sciences approaches with material science and physics principles. These approaches re leading us to develop strategies for improving plant productivity and stress tolerance. I will summarize our findings based on integrated approaches to show how science-based innovations can be developed to reduce water, chemical and space footprint and increase productivity and quality of produce in urban farming.

Invited Talk: Dissecting Cocoa Beans Fermentation through Metagenomics Analysis

Conference: Food Innovation and Engineering Asia Conference

Year: 2019

Proceeding: 2019 Food Innovation and Engineering (FOODIE) Asia Conference

Group: General Submissions

Session: Session 1: Technologies for Primary Production

Time: Thursday, December 5, 2019 - 12:00pm-12:25pm

Authors:

González Barrios, A. F., Universidad de los Andes
Fernandez Niño, M. A., Universidad de los Andes
Rodriguez Cubillos, M. J., Universidad de los Andes
Restrepo Restrepo, S., Universidad de los Andes
Cepeda Hernandez, M. L., Universidad de los Andes
Olarte, H., Casa Luker
Del Portillo, P., Corpogen
Anzola, J. M., Corpogen
Chica Morales, M. J., Casa Luker

Aguirre, J. L., Casa Luker
Calderón, D., Casa Luker
Rodriguez, C. M., Casa Luker

Cocoa cultivation in Colombia is highly relevant as Colombian cacao has unique organoleptic properties that have placed it into the category of special cocoa. One of the variables with the greatest impact is the presence and activity of the microorganisms associated with the fermentation processes, at which time most of the organoleptic properties of chocolate are acquired. Therefore, a comprehensive description of the microbial communities at the species level, associated with each day of a standardized process of fermentation, has not been reported yet. The objective of this research is to identify the species of microorganisms involved throughout a process of fermentation of fine aroma cocoa beans to produce chocolate and construct two genome-scale metabolic models (GSMMs), one for first day and other for the last one, to identify main metabolites related with this process. For this, a sample of the cocoa beans was extracted for each day during this process in a cocoa production farm. Sequencing of 16S rRNA and ITS2 was performed for the detection of bacteria and fungi by means of MiSeq Illumina. The identification allowed to describe that the dominant species are Bacillus subtilis, Lactobacillus fermentum, Saccharomyces cerevisiae, Acetobacter aceti, Fructobacillus pseudofilcunes and Halomonas stevensi. Interestingly, our study describes the presence of Pestalotiposis rhodondendri, Lysinibacillus xylanilyticus and Lactobacillus vini which had not been previously reported in cocoa beans fermentation. With GSMMs, the presence of sugars, alcohol and citric acid was register for first day and acetic acid production for the last day of fermentation process.

Keynote Talk: The Use of Microbiological Risk Assessment to Reduce Foodborne Disease Risk

Conference: Food Innovation and Engineering Asia Conference

Year: 2019

Proceeding: 2019 Food Innovation and Engineering (FOODIE) Asia Conference

Group: General Submissions

Session: Session 2: Food Safety

Time: Thursday, December 5, 2019 - 1:30pm-1:55pm

Authors:

Schlundt, J., Nanyang Technological University

Science-based decisions about policies aimed at preventing foodborne contamination and illness have become very important in general, recently primarily based on the principles of Risk analysis. Risk analysis is a concept and framework fostered by the World Health Organization (WHO) and the UN Food and Agricultural Organization (FAO) in the mid-nineties, to ensure that regulatory decisions about foods are science-based and transparent. Risk analysis comprises Risk assessment, Risk management and Risk communication.

Over the last decades emerging pathogens have caused new and increasing problems all over the world. Examples of these pathogens are Salmonella, Campylobacter, Yersinia, Enterohemorrhagic E. coli (EHEC) and Listeria monocytogenes. The international trade agreements: World Trade Organization (WTO) SPS agreement (Article 2) establishes that sanitary measures should be based on scientific principles. Risk analysis should be used to enhance protection of human health and minimize the incidence of food-borne disease through establishing realistic and achievable levels of control of food-borne hazards, and basing food safety policies on the practical application of the results of Risk assessment and Risk management.

While risk assessment of chemical food safety hazards has been performed internationally for more than 50 years, microbiological risk assessment is a relatively new scientific discipline. It is important for everyone involved in this area to understand the general principles behind these assessments as well as the potential benefits and uncertainties related to the models used and the estimates arrived at using microbiological risk assessment. This presentation will describe examples of quantitative microbiological risk assessment (MRA), a tool containing food chain models (â€˜farm-to-forkâ€™) combined with dose-response evaluations, finally resulting in outcome risk estimates in the final Risk Characterization of MRA

Invited Talk: The Alphabet Soup of Singapore Food Safety Research - A*STAR, IFCS, FRESH and SIFBI

Conference: Food Innovation and Engineering Asia Conference

Year: 2019

Proceeding: 2019 Food Innovation and Engineering (FOODIE) Asia Conference

Group: General Submissions

Session: Session 2: Food Safety

Time: Thursday, December 5, 2019 - 1:55pm-2:20pm

Authors:

Smith, B., A*STAR

My talk will give a brief overview of some of the current and developing programmes within Singapore aimed at supporting food safety research. Particular focus will be on the A*STAR Innovations in Food and Chemical Safety (IFCS) Programme and examples of different projects we are working on related to the assessment of food ingredients. I will also briefly introduce the joint A*STAR, SFA and NTU platform FRESH, the Future Ready Food Safety Hub, and some of the future opportunities in Singapore linked to both programmes

Invited Talk: Isolation of Selective Nanobodies Against Illegal Antibiotics Used in Food Supply Chains

FOOD INNOVATION AND ENGINEERING CONFERENCE

Conference: Food Innovation and Engineering Asia Conference

Year: 2019

Proceeding: 2019 Food Innovation and Engineering (FOODIE) Asia Conference

Group: General Submissions

Session: Session 2: Food Safety

Time: Thursday, December 5, 2019 - 2:20pm-2:45pm

Authors:

Swofford, C. A.
Sinskey, A. J., Massachusetts Institute of Technology
Nordeen, S. A., Massachusetts Institute of Technology
Schwartz, T. U., Massachusetts Institute of Technology

Economically motivated adulteration of food supply chains presents a major threat to public health and, more generally, on the public trust in the integrity of the food system. The predominant adulterants in food products are illegal antibiotics that are used to prevent bacterial infection of poultry livestock and aquaculture. Overuse of these antibiotics in food supply chains risk the unintentional creation of â€œsuperbugsâ€ that are totally drug resistant. Tools that allow rapid detection of antibiotics in food products could have a major and longstanding positive impact on local food systems. Current capabilities, however, include either expensive lab-based tests which are quantitative and slow or rapid kits with low accuracy, low throughput, and an inability to test for multiple adulterants. This situation leads to very sparse testing of food supply chains that cannot meet the monitoring need in areas with a large volume of transactions. In order to improve monitoring of food supply chains, we have developed single-domain antibodies, or â€œnanobodiesâ€, that are highly selective for an individual antibiotic. The small size of nanobodies allows for improved solubility, shelf-life, and functionalization to existing sensor technologies. As an example, we have developed a nanobody that binds to chloramphenicol with a dissociation constant of 50nM by using crystallography and molecular dynamic simulations to improve its binding affinity. This paradigm can be used to develop additional nanobodies that target small molecule adulterants of interest in food supply chains with the goal of creating an array of recognition elements for sensor-based applications.

Invited Talk: Harnessing Technology, Fighting Food Fraud and Transforming Food Manufacturing

FOODIE Asia
FOOD INNOVATION AND ENGINEERING CONFERENCE

Conference: Food Innovation and Engineering Asia Conference

Year: 2019

Proceeding: 2019 Food Innovation and Engineering (FOODIE) Asia Conference

Group: General Submissions

Session: Session 2: Food Safety

Time: Thursday, December 5, 2019 - 2:45pm-3:10pm

Authors:

Lai, A., ProfilePrint

Thereâ€™s a name for it: food fraud. A number of high profile scandals in recent years, such as rice made with plastics, or cheap teas masquerading as expensive brews, have made people cautious about what they eat. To protect consumers from fraud and health issues caused by consuming these products, governments are relying on blockchain solution, QR codes on packages or sending samples to labs for testing â€“ but our scientific approach measures the food itself to prove what it claims to be, even when the packaging is removed. Recent technological advances in data analytics, machine learning and spectrometry have made what was once previously impossible, possible. Science is now able to identify food molecular structure cheaply and accurately - and that has had impressive impact on food manufacturing transparency and efficiency . The speaker will share more on how technology should be harnessed to improve the supply chain efficiency of the food industry, and more importantly strengthen the trust amongst consumers as we move into the new era of increasing food fraud and unscrupulous adulteration.

Plenary Talk: The Gut Microbiota in Early Life, Target for Nutrition and Health

Conference: Food Innovation and Engineering Asia Conference

Year: 2019

Proceeding: 2019 Food Innovation and Engineering (FOODIE) Asia Conference

Group: General Submissions

Session: Plenary Session II

Time: Friday, December 6, 2019 - 8:30am-8:55am

Authors:

Knol, J., Wageningen University

Our intestinal microbiota forms an integral part of normal human physiology, and disturbances of the normal gut microbiology have been linked to some important health and disease issues. Newborns are essentially sterile and the microbiota establishes from the very first days of life. These first colonizers may play an important role in the development of the symbiosis and may impact the long-term composition and activity of the microbiota. Considering the importance of the microbiota on the human immune, metabolic, and neurological systems, it is important to understand the dynamics and driving determinants of this development. The development of the human gut microbiota is dependent on many factors, as for example, host genetics, mode of delivery, antibiotic use, etc., but also the diet is clearly an important factor.

Exclusive human milk feeding is widely recommended as the first choice of infant nutrition, not only providing optimal nutrition, but also bioactive components that are crucial for optimal gut development, immune maturation, metabolic development, and even cognitive development. Since breast feeding may not always be possible, human milk substitutes have to provide the best nutrition and functional properties similar to human milk. Based on insights from human milk research and an increasing understanding of the importance of the gut microbiota, specific concepts for infant formulas have been developed in the past years. The impact of these on the early microbiota as well as some potential health benefits will be presented.

Keynote Talk: Gentle Food Processing for Better Sustainability

FOOD INNOVATION AND ENGINEERING CONFERENCE

Conference: Food Innovation and Engineering Asia Conference

Year: 2019

Proceeding: 2019 Food Innovation and Engineering (FOODIE) Asia Conference

Group: General Submissions

Session: Session 3: Sustainability of Food Production

Time: Friday, December 6, 2019 - 9:30am-9:45am

Authors:

Boom, R. M., Wageningen University and Research Center

It is evident that the availability of sufficient, high quality food without over-exploiting our Earth, will be our prime challenge in the coming decades. This challenge is complicated by an already changing climate, which reduces the potential to intensify the cultivation of the current agricultural land, and the potential to sustainably cultivate new land, and may complicate the use of the oceans for food production. It is interesting that the system that converts our raw materials on the land, into edible foods, is quite inefficient. Of the total amount of protein that is grown on the land, less than 6% reaches our plates (Alexander et al., 2016). This is due to incomplete harvesting, losses in processing steps, but also due to the conversion of plant proteins into animal products. Making the overall food system more efficient could easily help us meet the targets set for 2050, without having to produce more primary raw materials. Our current generation of plant based protein foods is however based on an efficient method for the isolation of plant proteins, having low yield and requiring significant amounts of water, chemicals and energy. This strongly reduces the potential of these products, to improve the overall conversion efficiency, and thus to achieve our food production goals for 2050. This can change if we change our perception of ingredient production. By not refining our ingredients to purity, but to limit the refining, one can create intermediates that have excellent (techno)functionality, i.e. leading to well-structured and stable foods, but can also have better nutritional value, due to the retention of fibre and micro-nutrients. At the same time, the raw material is better utilised, leading to less side streams, while their quality remains much higher. The presentation will discuss the principle of gentle processing, present some typical intermediates that can be obtained, and will discuss their properties and uses for our future food. Overall, the concept of gentle processing will allow us to achieve more food from less raw materials, that fits in a healthier diet, and will result in significantly less waste.

Invited Talk: Digitalization and AI in Urban Aquaculture

FOODIE Asia
FOOD INNOVATION AND ENGINEERING CONFERENCE

Conference: Food Innovation and Engineering Asia Conference

Year: 2019

Proceeding: 2019 Food Innovation and Engineering (FOODIE) Asia Conference

Group: General Submissions

Session: Session 3: Sustainability of Food Production

Time: Friday, December 6, 2019 - 9:45am-10:00am

Authors:

Shishehchian, F., Blue Aqua International

The worldâ€™s demand for seafood has tremendously outweighed the Earthâ€™s natural ability to replenish its fish stocks, with many areas experiencing overfishing. Therefore, the aquaculture industry seeks new technology to increase production and sustainability of seafood. The use of artificial intelligence in aquaculture for intensification has been explored and discussed with relevance to current aquaculture issues, to rapidly produce high fish stocks with low cost and labour.

Current technologies in aquaculture have proven that cognitive automation contributes hugely to reduce labor, time and utility costs. These computerised systems have increased efficiency, conserve our precious resources and improved physiological rates (e.g. growth) of cultured species. In the future, the succession of Artificial Intelligence (AI) in aquaculture would greatly enhance processes in farms, increasing its production and profits in the long-term. This informative sharing session will provide insights of AI and its integration into systems - feed control, water parameters monitoring, biological filtration and removal and ultimately, applications in Smart-Farms for intensification. The intensive culture of high-value species will also be discussed with relevance to how the addition of AI in these cultures will significantly aid in the diminishing population and escalating demand.

Mexican Foods: What Can be Done to Preserve Traditions, Health and the Environment?

Conference: Food Innovation and Engineering Asia Conference

Year: 2019

Proceeding: 2019 Food Innovation and Engineering (FOODIE) Asia Conference

Group: General Submissions

Session: Session 3: Sustainability of Food Production

Time: Friday, December 6, 2019 - 10:00am-10:15am

Authors:

Gutierrez, J., Tecnológico de Monterrey

Mexican food is an interesting example of the blend of two cultures. One of the most outstanding food processing of our pre-hispanic ancestors is maize lime cooking or nixtamalization. Economic, social and environmental sustainability criteria such as commercial opportunities, use of materials and employment creation are very important factors that should be considered as indicators of sustainability. Nixtamalization generates a waste stream that contains phytochemicals that can be extracted to generate active principles such as ferulic acid derivatives. Additionally, it is a fiber source and calcium that can be used to enrich bakery products without detrimental effects on color of flavor. Ferulic acid is one of the bioactive compounds that can be recovered from this waste stream and converted into different conjugated forms to improve its functionality.

Invited Talk: Sustainability and the Development of Food Production Systems

Conference: Food Innovation and Engineering Asia Conference

Year: 2019

Proceeding: 2019 Food Innovation and Engineering (FOODIE) Asia Conference

Group: General Submissions

Session: Session 3: Sustainability of Food Production

Time: Friday, December 6, 2019 - 10:15am-10:30am

Authors:

Wuertz, S., Nanyang Tehcnological University

Providing and maintaining adequate supplies of food and water has always been essential to human settlements, regardless of size and state of development. In todayâ€™s world all societies are faced with environmental and political uncertainties that will require novel approaches to how we produce food. Aspects of sustainability include both recycling of water during production (thereby reducing discharge of used water) and potable water reuse, where wastewater is treated to high standards that satisfy drinking water regulations. In the future, reclaimed and desalinated water will become a mainstay of water supplies in many geographic regions. Indeed, in some countries highly treated wastewater is already reused for agriculture (for example, about 90% in Israel and 30% in Spain) and the will to implement such solutions is driven by economic necessity.

Considering the above, innovative food production systems that go beyond traditional forms of agriculture are faced with a dual challenge: (i) how to procure water supplies that are reliable, safe and affordable and (ii) how to design systems that minimize water usage without affecting quality and yield. Given the worldwide growth in resource consumption and waste generation with increasing population and urbanization, the concept of a circular economy is a high-level strategy to move away from the dominant â€œtake-make-disposeâ€ (â€œonce throughâ€) way of economic development, and achieve a shift towards environmental sustainability. Waste management is one of the significant aspects of the circular economy, in which bio-waste serves as a valuable organic resource. Renewable carbon can be extracted or transformed into diverse end-products such as food, feed, bio-based chemicals, biopolymers, fuels and bioenergy. Recycling of bio-wastes contributes significantly to circular economy objectives and sustainable development; moving towards a circular bio-economy enhances competitiveness, positive economic development and job creation by organizational, social and technological innovation. Besides the development of technical innovations, appropriate decisions and strategies regarding sustainability and economic feasibility have to be made.

Keynote Talk: The Intersection of Diet, Brain, Gut Microbiome and Behaviour - Intervening to Improve Cognition

Conference: Food Innovation and Engineering Asia Conference

Year: 2019

Proceeding: 2019 Food Innovation and Engineering (FOODIE) Asia Conference

Group: General Submissions

Session: Session 4: Health and the Gut microbiome

Time: Friday, December 6, 2019 - 12:30pm-12:55pm

Authors:

Morris, M., UNSW Sydney

Our research attempts to model the western lifestyle, investigating how provision of varied, energy rich diets can override the control mechanisms that should maintain body weight. Rats fed an obesogenic, cafeteria style diet consistently show deficits in hippocampal-dependent memory tasks, and reduced diversity of their gut microbiome compared to control rats. Such behavioural deficits were independent of weight differences, as rats consuming diets high in saturated fat or high in sugar, for just two weeks, had impaired spatial memory even in the absence of caloric excess. We found that the memory deficits were associated with changes in the gut microbiota composition and genes related to inflammation in the hippocampus, a key brain region for memory and learning.

More recently we have investigated whether the bacteriostatic antibiotic, minocycline, which is reported to exert anti-inflammatory effects, can modulate spatial memory. Again, cafeteria diet produced persistent deficits in spatial memory (novel place recognition) that were prevented by minocycline cotreatment. Of interest, chow fed rats treated with minocycline performed worse than those treated with vehicle. Faecal microbiota alpha diversity was reduced by both cafeteria diet and minocycline treatment, but these reductions were not associated with performance on the novel place task. However, abundances of specific OTUs within *Bacteroides* and *Lactobacillus* were associated with place task performance. Together, studies such as these suggest the gut microbiota could play a causal role in regulating behaviour. Current experiments are exploring the impact of fecal transfer on memory performance in rats consuming the obesogenic diet.

Invited Talk: Metabolomics Detects Functions of Gut Microbes-Functional Microbiomes

FOODIE Asia

FOOD INNOVATION AND ENGINEERING CONFERENCE

Conference: Food Innovation and Engineering Asia Conference

Year: 2019

Proceeding: 2019 Food Innovation and Engineering (FOODIE) Asia Conference

Group: General Submissions

Session: Session 4: Health and the Gut microbiome

Time: Friday, December 6, 2019 - 12:55pm-1:20pm

Authors:

Yulan, W., Lee Kong Chian School of Medicine (LKCMedicine)

In recent years, the gut microbiota has gained increasing attention not only due to its impact on host physiology and immune system development, but also because perturbation of the microbiota may result in a number of diseases, such as diabetes, infectious disease, cardiovascular disease as well as cancer. Thus, understanding the impact that the microbiota plays in various diseases and the way microbes can be modified are fundamental for the development of novel therapeutic strategies for disease treatment and disease prevention. One way of investigating the changes gut microbes is to measure then using metagenomics tools. However, many of genes in microbes world have not been annotated and their functions are unknown. Metabolomics provides alternative way to detect functions of perturbed microbiome communities-functional microbiomes. In the current presentation, examples of metabonomics investigation on how dietary shapes gut microbes and how the changes are reflected in changes of metabolomes of host will be given.

Invited Talk: Early Life Microbiome, A Risk Factor for Pediatric Allergy

Conference: Food Innovation and Engineering Asia Conference

Year: 2019

Proceeding: 2019 Food Innovation and Engineering (FOODIE) Asia Conference

Group: General Submissions

Session: Session 4: Health and the Gut microbiome

Time: Friday, December 6, 2019 - 1:20pm-1:45pm

Authors:

Lay, C., Danone Nutricia Research

The first 1000 days of life is recognized as an important window to nurture child health and development, and increasing body of evidence indicate that a compromised microbiome is a risk factor for the development of non-communicable diseases. The human gut microbiome is transmitted from one generation to the next. This transgenerational microbial inheritance occurs in early life. Such vertical transmission contributes to educate our immature immune, metabolic and neurocognitive systems during foetal and early life, highlighting the role of the microbiome in health and diseases. Human milk plays a key role in nurturing a healthy microbiome in early life, and maintaining overall health. Several epidemiological studies have depicted associations of prenatal/ postnatal exposure to antibiotics and C-section births with the development of non-communicable diseases such as allergies and obesity. It appears therefore that a disruption of the vertical transmission and early microbial colonization may have long-term consequences on the health of the offspring. Nutritional intervention could be harnessed as a mean to reduce the disease risk associated with a compromised microbiome.

Invited Talk: VALUE Addition and Influence of Succilynation Levels on the Quality of Discorea Villosa Starch

FOODIE Asia
FOOD INNOVATION AND ENGINEERING CONFERENCE

Conference: Food Innovation and Engineering Asia Conference

Year: 2019

Proceeding: 2019 Food Innovation and Engineering (FOODIE) Asia Conference

Group: General Submissions

Session: Session 4: Health and the Gut microbiome

Time: Friday, December 6, 2019 - 2:10pm-2:35pm

Authors:

Raji, A., Kwara State University, Malete
Ismael, S., Kwara State University, Malete

Recent increase in demand for starch in food applications has called for research into other new starch sources. *Dioscorea villosa* tuber is an underutilized starch source and its starch was isolated and succilynated at 0, 3, 6, 9 and 12% (0-12% succinic anhydride: 88-100% starch) using standard procedures. The degree of substitution, the chemical composition, functional properties, anti-nutritional factors, pasting properties and the atomic spectra of the samples were determined using standard methods. Data obtained were analysed using ANOVA. The percentage succinyl and the degree of substitution increased from 0 - 10.45% and 0- 0.19% respectively. The amylase and amylopectin contents varied from 49.88-59.70% and 40.29-50.11% respectively. The phytate, tannin, oxalate and saponin ranged from 0.08 - 0.13 mg/g, 0.04 - 0.08mg/g, 0.21 - 0.31mg/g and 0.18 - 0.26mg/g respectively. The peak viscosity, trough, break down, final viscosity, set back, peak time and pasting temperature values ranged from94.32-274.69RVA, 83.50-197.50RVA, 7.94-81.28RVA, 113.69-367.39RVA, 30.14-173.89RVA, 5.30-6.94mins, and 84.78-94.70°C respectively. Isolation and subsequent succilynation improved the wild yam starch quality and reduced its anti-nutritional factors. Succilynation at 9% had the best functional and pasting properties, therefore making it suitable as a potential emulsifier and stabilizer in food applications.

Keywords: Wild yam starch, succilynation, functional properties, degree of substitution

Keynote Talk: Sugar and Salt - A Fresh Look at Well-Known Ingredients

FOOD INNOVATION AND ENGINEERING CONFERENCE

Conference: Food Innovation and Engineering Asia Conference

Year: 2019

Proceeding: 2019 Food Innovation and Engineering (FOODIE) Asia Conference

Group: General Submissions

Session: Session 5: Future Foods, Innovations and Novel Nutritional Solutions

Time: Friday, December 6, 2019 - 2:50pm-3:15pm

Authors:

Oertling, H., Nestlé Health Science

Ionic cocrystals have recently gained attention as multi-component pharmaceutical materials capable of improving the physicochemical properties of the respective pharmaceutically active compound. Contrastingly, ionic cocrystals between common carbohydrates and alkali or alkaline earth halides have been investigated since the beginning of chemistry and most notably in the context of food science. Throughout several centuries, these compounds have been inspected with the specific perspective of contemporary interest. In most cases, they can conveniently be prepared from the ionic salt and the respective neutral carbohydrate via direct solution crystallization and therefore offer an opportunity for fundamental studies in the fields of crystal engineering, bridging food science and solid-state chemistry. Certainly ionic cocrystals remain an underexplored class of compounds and this talk gives an overview of the work that has been performed until today, hoping to elicit further contributions in the discipline.

Invited Talk: Sensory Influences on Oral Processing & Energy Intake: Challenges and Opportunities

FOODIE Asia
FOOD INNOVATION AND ENGINEERING CONFERENCE

Conference: Food Innovation and Engineering Asia Conference

Year: 2019

Proceeding: 2019 Food Innovation and Engineering (FOODIE) Asia Conference

Group: General Submissions

Session: Session 5: Future Foods, Innovations and Novel Nutritional Solutions

Time: Friday, December 6, 2019 - 3:15pm-3:40pm

Authors:

Forde, C., A*STAR, Singapore

When it comes to human energy intake not all calories are equal as a foods sensory cues can strongly influence the extent to which it is consumed (McCrickerd & Forde, 2016). Food texture, taste and aroma are influential before and during meals, and are used to direct food choice and guide the energy consumed to fullness. We are interested in understanding how perceptual responses to food can moderate eating behaviours and energy intake. When energy dense foods are consumed at a faster eating rate they promote greater energy intakes and deliver poorer satiety *per* kcal. In a recent study, participants consumed >500kcals *per* day when consuming a diet of ultra-processed foods, and these higher intakes were associated with faster energy intake rates (Hall et al 2019). Our research in children and adults has shown eating rate to be a consistent behaviour that is positively associated with increased energy intake and adiposity (Forde, Fogel and McCrickerd 2019). Epidemiological studies have consistently shown that individuals that self-report a higher eating rate are at greater risk of overweight and obesity (Teo and Forde 2019) and we have recently confirmed this in a population based sample of adults in Singapore. To counter this, we have shown that it is possible to manipulate a foods texture to reduce the rate of energy intake and decrease meal size. Importantly these reductions in energy intake can be achieved with sustained hedonic appeal and post-meal satisfaction. Taken together, these findings highlight an opportunity to apply sensory manipulations and changes to food textures to promote oral processing behaviour that support reduced energy intake, while sustaining satisfaction. Addressing the serious public health challenges posed by the modern food environment will require changes in food formulation and intake behaviour. Food sensory properties can support healthier food choices and intakes and inform the development of successful behavioural and dietary strategies for better management of chronic conditions such as obesity and type-2 diabetes.

References:

1. McCrickerd, K. and Forde, C.G. (2016). Sensory influences on food intake control: moving beyond palatability. *Obesity Reviews,* Volume 17, 18â€“29. DOI: 10.1111/obr.12340.
2. Forde, C. G., Fogel, A., & McCrickerd, K. (2019). Children's Eating Behaivors and Energy Intake: Overlapping Influences and Opportunities for Intervention. In *Nestle Nutrition Institute*Workshop series (Vol. 91, pp. 55-67) Karger Press. DOI:10.1159/000493695.
3. Teo P.S., Forde C.G. (2020) The Impact of Eating Rate on Energy Intake, Body Composition and Health. In: Meiselman H. L. (Ed.)*Handbook of Eating and Drinking.* Springer, pp 1-27,https://doi.org/10.1007/978-3-319-75388-1.

Invited Talk: The Use of Novel Fractionation and Isolation Technologies for the Recovery of Biomolecules with Application in Food Formulations

FOODIE Asia
FOOD INNOVATION AND ENGINEERING CONFERENCE

Conference: Food Innovation and Engineering Asia Conference

Year: 2019

Proceeding: 2019 Food Innovation and Engineering (FOODIE) Asia Conference

Group: General Submissions

Session: Session 5: Future Foods, Innovations and Novel Nutritional Solutions

Time: Friday, December 6, 2019 - 3:40pm-4:05pm

Authors:

Benavides, J., Tecnológico de Monterrey

In accordance with the current trends in bioprocessing each day new extraction, recovery and purification techniques are design and applied for the isolation of biomolecules with different applications. In this regard, many of those biomolecules have application in the food industry as antimicrobials, antioxidants, structuring proteins, pigments, etc. Therefore, the application of these novel extraction, recovery and purification technique have a significant role in the development of new food products and food processes. One of the techniques with more potential for the extraction, primary recovery and partial purification are the Aqueous Two-Phase Systems (ATPS). These liquid-liquid fractionation systems have been characterized for the downstream processing of low molecular weight compounds, biomacromolecules, bioparticles and even whole cells. ATPS allow to achieve process integration and intensification, while presenting a biocompatible environment for most bioactives. To exemplify the potential of ATPS the development of a pilot-plant scale process for the recovery of B-Phycoerythrin (BPE) from *Porphyridium cruentum* is presented as example.

Invited Talk: Role of Food Technology and Future Foods in Enhancing Human Health

Conference: Food Innovation and Engineering Asia Conference

Year: 2019

Proceeding: 2019 Food Innovation and Engineering (FOODIE) Asia Conference

Group: General Submissions

Session: Session 5: Future Foods, Innovations and Novel Nutritional Solutions

Time: Friday, December 6, 2019 - 4:05pm-4:30pm

Authors:

Jeyakumar, H., National University of Singapore

It is hard to imagine that the science of food and nutrition is merely a hundred years old. The discovery of vitamin in 1914 is conventionally viewed as the date of inception of the science of nutrition. During the intervening decades, numerous advances in Food & Nutrition have enabled us to understand the role of food on health outcomes. Advances in food technology has played a major contribution to societyâ€™s health. Through better food preservation, a greater variety of nutritious products and food fortification. However, current progress in our understanding of the interplay between health, diet and nutrition is opening up exciting opportunities to use foods to deliver not just improved nutrition but importantly a better quality of life for all consumers. This presentation will examine the evolving link between food technology and nutrition from an historical and scientific perspective. At the dawn of civilization, hunter-gathererâ€™s survival was paramount and food was secured by foraging several hundred plants & animal sources. Today, less than 20 staple foods are used to feed the worldâ€™s population. Food technology has provided the technical ability to fashion foods with unique nutritional, sensory and gustatory attributes. The future progress in Food technology and nutrition research can only be achieved if private-public partnership is nurtured. Given the unique ecology and environment of Asia, it is important that researchers in this region craft their research questions in relation to the contemporary needs of this region.

Invited Talk: Agave: A New Age of Functional Foods

Conference: Food Innovation and Engineering Asia Conference

Year: 2019

Proceeding: 2019 Food Innovation and Engineering (FOODIE) Asia Conference

Group: General Submissions

Session: Session 5: Future Foods, Innovations and Novel Nutritional Solutions

Time: Friday, December 6, 2019 - 4:30pm-4:55pm

Authors:

Alejandra, J., Tecnológico de Monterrey

Agave sap concentrate (ASC) is an artisanal Mexican food produced by evaporation of fresh agave sap, also known as â€œaguamielâ€. The concentration process is conducted immediately after harvesting the fresh sap to avoid fermentation by the endogenous microbiota. Agave sap is obtained from selected agave species such as *Agave americana*, *A. salmiana* and *A. atrovirens*. This artisanal product has gained popularity due to its antioxidant capacity, phytochemical content, as well as anticancer and antidiabetic properties. Among the phytochemicals related to the bioactivities, saponins are the responsible for the effects on the regulation of cancer cell growth and regulation of carbohydrate and fat metabolism. But the concentration of these compounds depends on the region from where the sap is collected and it has been shown that endogenous microbiota can also affect the saponin profile in agave sap. Therefore the isolation of microorganisms able to produce fermented products with a selected profile of saponins is under investigation for a new generation of nutraceuticals.

Antibacterial and Biofilm-Debriding Hydrogel Wound Dressing

Conference: Food Innovation and Engineering Asia Conference

Year: 2019

Proceeding: 2019 Food Innovation and Engineering (FOODIE) Asia Conference

Group: Poster Session

Session: Poster Session

Authors:

Yeo, C. K., Nanyang Technological University

Multidrug resistance and the refractory character of bacterial biofilms are among the most difficult challenges in infection treatment. Current antimicrobial strategies typically are much more effective for prevention of biofilm formation than for eradication of established biofilms; these strategies also leave dead bacteria and endotoxin in the infection site, which impairs healing. We report a novel hydrogel that eradicates biofilm bacteria by non-leaching-based debridement followed by ex situ contact-killing (DESCK) away from the infection site. The debridement effect is likely due to the high water swellability and microporosity of the cross-linked network which is made from polyethylene glycol dimethacrylate tethered with a dangling polyethylenimine (PEI) star copolymer. The large pore size of the hydrogel makes the cationic pore walls highly accessible to bacteria. The hydrogel also degrades in the presence of infection cells, releasing star cationic PEI into the infection site to contact-kill bacteria remaining there. DESCK hydrogel effectively kills (>99.9% reduction) biofilms of methicillin-resistant Staphylococcus aureus (MRSA) and carbapenem-resistant Pseudomonas aeruginosa (CR-PA) and Acinetobacter baumannii in a murine excisional wound infection model. Silver-based wound dressings (controls) showed almost no killing of CR-PA and MRSA biofilms. This DESCK hydrogel greatly reduces the bioburden and inflammation and promotes wound healing. It has great potential for diverse infection treatment applications.

Block Copolymer Nanoparticles Remove Biofilms of Drug-Resistant Gram-Positive Bacteria By Nanoscale Bacterial Debridement

FOODIE Asia

FOOD INNOVATION AND ENGINEERING CONFERENCE

Conference: Food Innovation and Engineering Asia Conference

Year: 2019

Proceeding: 2019 Food Innovation and Engineering (FOODIE) Asia Conference

Group: Poster Session

Session: Poster Session

Authors:

Li, J., NTU

Bacterial infection has become the main threat to human, especially the biofilm associated infections. Once developed into biofilm form, it will become more resistant to the conventional antibiotics. Hence, many antibiotics, natural antimicrobial peptides and synthetic antimicrobial agents have been studied as antibiofilm agents. However, they are usually suffered from the problem of toxicity and limited life span. Therefore, a novel antibiofilm cationic copolymer has been synthesized to study both the antibacterial and antibiofilm ability. Interestingly, the polysaccharide-based amphiphilic copolymer can self-assembled into nanoparticle form which donâ€™t have any antibacterial effect but excellent preformed biofilm removal ability. Within the cationic charge and hydrophobic segments hidden inside the core of nanoparticles, the bacterial killing ability became disappeared; however, the antifouling shell of the polysaccharide as well as the nanoparticles form enhanced the biofilm dispersal ability by the mechanism we called â€œnanoscale bacterial debridementâ€. Our in vitro results showed that the polymeric nanoparticles have the antibiofilm ability towards several multi-drug resistant and clinically relevant strains, with the efficacy much higher and/or similar to the conventional standard antibiotics. In vivo data also showed that such nanoparticles have MRSA biofilm removal efficacy higher than vancomycin. Further, both the in vitro and in vivo data showed our nanoparticles have good biocompatibility with low hemolysis and cytotoxicity. Overall, this novel biofilm removal approach provides exciting opportunities for treatment of multi-drug resistant biofilm infections and which further may have widespread applications.

Cationic Peptidopolysaccharide Shows Excellent Antimicrobial Efficacy with Good Biocompatibility

Conference: Food Innovation and Engineering Asia Conference

Year: 2019

Proceeding: 2019 Food Innovation and Engineering (FOODIE) Asia Conference

Group: Poster Session

Session: Poster Session

Authors:

Zheng, H., Nanyang Technological University Singapore

Cationic antimicrobial peptides (AMPs) and polymers are active against many multi-drug resistant (MDR) bacteria but only a limited number of these compounds are in clinical use due to their unselective toxicity. The typical strategy for achieving selective antibacterial efficacy with low mammalian cell toxicity is through balancing the ratio of cationicity to hydrophobicity. Herein, we report a cationic nanoparticle self-assembled from chitosan-graft-oligolysine (CSM5-K5) chains with ultra-low molecular weight (1450 Daltons) that selectively kills bacteria. Further, hydrogen bonding rather than the typical hydrophobic interaction causes the polymer chains to be aggregated together in water into small nanoparticles (with about 37nm hydrodynamic radius) to concentrate the cationic charge of the lysine. When complexed with bacterial membrane, these cationic nanoparticles synergistically cluster anionic membrane lipids and produce greater membrane perturbation and antibacterial effect than would be achievable by the same quantity of charge if dispersed in individual copolymer molecules in solution.

Comparative Fecal Metabolomics Analysis of Healthy Elderly and Young Male Singaporean

FOODIE Asia
FOOD INNOVATION AND ENGINEERING CONFERENCE

Conference: Food Innovation and Engineering Asia Conference

Year: 2019

Proceeding: 2019 Food Innovation and Engineering (FOODIE) Asia Conference

Group: Poster Session

Session: Poster Session

Authors:

Chen, L., NANYANG TECHNOLOGICAL UNIVERSITY
Teh, J., NANYANG TECHNOLOGICAL UNIVERSITY
Schlundt, J., Nanyang Technological University
Conway, P., NANYANG TECHNOLOGICAL UNIVERSITY

In recent years, emerging evidence from gut metagenomics studies has demonstrated that gut microbiome has a crucial role in health and aging. The gut metabolome, which is generated by the gut microbiota and host, is the link between the gut microbiome and host. Thus, the gut metabolomics analysis can provide deep insights into the relationship between the gut microbiome and human health. In this study, the GC-MS based metabolomics platform was used to perform a comparative metabolomics analysis of faeces samples from 33 healthy elderly and 32 young male Singaporeans.

The observed metabolite profiles showed significantly different levels (FDR value<0.05) between young and elderly male individuals. Elderly faecal samples contained higher relative concentrations of long chain fatty acids, but contained lower levels of short chain fatty acids; lower levels of branch chain amino acids and other amino acids and metabolites involved in amino acid metabolism; and lower levels of metabolites involved in carbohydrate metabolism. The observed significantly changed metabolites datasets suggests age probably contributes to fatty acid metabolism, amino acid metabolism and carbohydrate metabolism alterations. This work not only provides an important reference dataset for understanding the concentrations and variation in the Asian faecal metabolome, but also adds to the growing evidence of the importance of fatty acids and amino acids in chronic metabolic disease progression. In future, the combination of metabolomics, metagenomics and transcriptomics will enable more detailed studies of these biological pathways and the changes with age and provide targeted gut microbiome management strategies that promote healthy ageing.

Deep-Ultraviolet Light Emitting Chips for Food Safety

FOOD INNOVATION AND ENGINEERING CONFERENCE

Conference: Food Innovation and Engineering Asia Conference

Year: 2019

Proceeding: 2019 Food Innovation and Engineering (FOODIE) Asia Conference

Group: Poster Session

Session: Poster Session

Authors:

Sharma, V. K., Nanyang Technological University
Tan, S. T., Nanyang Technological University
Haiyang, Z., Nanyang Technological University
Shendre, S., Nanyang Technological University
Baum, A., LightLab Sweden AB
Chalvet, F., LightLab Sweden AB
Tirén, J., LightLab Sweden AB
Demir, H. V., Nanyang Technological University

In the food industry, the common techniques used to disinfect foods are mostly categorized into thermal and non-thermal processes. Thermal processes are known to affect the nutritional and sensory characteristics of the food. Therefore, non-thermal processing techniques i.e., ultrasonic, pulsed electric fields, high pressure, and UV light are getting more popular. Out of these techniques, UVC disinfection (250-280 nm) using mercury lamps is an emerging technology for disinfection in the food industry since it cannot be overdosed, does not generate by-products and possess no safety hazards. UVC disinfection is a physical process that kills bacteria, viruses, and other microbes by directly disrupting the molecular bonds that hold their DNA together. Theses UVC exposed microbes completely lose their reproductivity and are destroyed completely. Interestingly, this is still not a popular disinfection technique used in food industries because of the issues related to the safety (presence of toxic mercury), high cost and low penetration depth of UV light in a different physical medium.

To overcome these issues, we have developed a novel mercury-free deep-UV light chips based on on-chip cathodoluminescence having an emission spectrum coinciding with the germicidal effectiveness curve. Our deep-UV chips offer ultrahigh germicidal reduction, immediate start-up, low maintenance, instant ON/OFF capability, high operational lifetimes, work efficiently both at low and high temperatures and are smaller in size providing complete design freedom. We believe our technology will be promising for the disinfection in food industries by treating the food contact surfaces and by extending the shelf-life of fresh produce.

Do Dogs like Insects? a Palatability Test of Insect-Protein Formulated Dog Food

Conference: Food Innovation and Engineering Asia Conference

Year: 2019

Proceeding: 2019 Food Innovation and Engineering (FOODIE) Asia Conference

Group: Poster Session

Session: Poster Session

Authors:

Wein, L., Protenga
Yusnaieny, S., Protenga
Singam, E., Wondercare Veterinary Klinik
Baskaran, G., Wondercare Veterinary Klinik
Law, Y., Protenga
R, P., Subang Jaya Medical Centre

Meat and fish as protein in dog food has a significant environmental footprint (900m pet dogs globally). Insects, particularly Black Soldier Fly Larvae (BSF, Hermetia Illucens), could be a more sustainable protein alternative. A study at the University of Veterinary Medicine

Hannover, Germany (Meyer et al., 2019) demonstrated similarity between digestibility of BSF meal and poultry meal in dogs (83.4% vs 83.1%) and revealed a â€œpositive effect on faecal quality.â€

Palatability is another success factor for a new ingredient. We formulated an AAFCO-compliant extruded diet with BSF meal as the only animal protein source (40%, no palatant or oil coating) and compared its palatability to an equivalent formula using fishmeal instead of BSF, a premium and a low-cost commercial diet (â€œHillâ€™s Science Dietâ€ and â€œIdolâ€ Lucky Star Brand). Nine healthy adult dogs were selected, familiarised with all four diets over one week and tested individually for three consecutive feedings in their normal environment. For each session, dogs were offered small amounts of two of the diets (BSF vs each of the three controls), introduced by sniffing and then allowed to choose freely. First bite and finishing order were observed.

Results show clear preference for BSF over fishmeal (75% tried and finished BSF first). Initial attractiveness of Premium was higher (63%), but finishing rate was equivalent (82%). Compared to low-cost, BSF was slightly preferred (52%) and more dogs finished the BSF diet (89% vs 78%). Overall, our results suggest good palatability of insect protein-based diets for dogs.

Effect of Plasma Treatment (He/CH4) on Glass Surface for the Reduction of Powder Flux Adhesion in the Spray Drying Process

FOOD INNOVATION AND ENGINEERING CONFERENCE

Conference: Food Innovation and Engineering Asia Conference

Year: 2019

Proceeding: 2019 Food Innovation and Engineering (FOODIE) Asia Conference

Group: Poster Session

Session: Poster Session

Authors:

Zubairi, S., UNIVERSITI KEBANGSAAN MALAYSIA
Ramlan, N., Universiti Kebangsaan Malaysia (UKM)

A 50Hz glow discharge He/CH_4 plasma was generated and applied for the modification of glass surface to reduce powder adhesion on the wall of spray dryer. The hydrophobicity of the glass samples determined by the water droplet contact angle and adhesion weight on glass, dependent on the CH_4 flow rate and plasma exposure time. There was a peak that appeared at 1470 cm^{-1} on the surface of treated glass indicating the presence of CH_3 groups from ATR-FTIR data. Surface morphology analysis using scanning electron microscopy (SEM) showed changes of roughness in the surface-treated glass. The presence of alkyl group (CH_3) that deposited on the glass surface is one of the factors that contribute to the increase in the surface roughness. The surface roughness will reflect the value of contact angle where hydrophobic surface are rougher compared to hydrophilic surface. The plasma treatment could enhance the value of the contact angle and thus reduced the adhesion on the spray dryer glass surface.

Effects of Trimethylamine N-Oxide on Inflammation and the Gut Microbiota Alteration Induced By Helicobacter Pylori

FOOD INNOVATION AND ENGINEERING CONFERENCE

Conference: Food Innovation and Engineering Asia Conference

Year: 2019

Proceeding: 2019 Food Innovation and Engineering (FOODIE) Asia Conference

Group: Poster Session

Session: Poster Session

Authors:

Wu, D., Sichuan univesity
Cao, M., Core Laboratory, School of Medicine, Sichuan Provincial People's Hospital Affiliated to University of Electronic Science and Technology of China, Chengdu, 610072, P.R. China
Zhao, J., Key Laboratory of Biological Resource and Ecological Environment of Chinese Education Ministry, College of Life Sciences, Sichuan University

Diet is one of the factors contributing to symptom of *Helicobacter pylori (H. pylori)* infection. Trimethylamine N-oxide (TMAO), a diet-related microbial metabolite, is associated with inflammatory and metabolic diseases. The aim of this study is to investigate the effects of

TMAO on inflammation and gut microbiota composition in *H. pylori*-infected mice via biochemical analyses, gene expression profiling and 16S rRNA sequencing. The *in vitro* experiments showed that TMAO not only increased the expression of growth- and metabolism-associated genes and the urease activity of *H. pylori*, but also elevated the production of virulence factors. Moreover, TMAO altered the immune response and promoted expression of IL-6, CXCL1 of GES-1 cells with *H. pylori*infection. Further *in vivo* experiments showed that TMAO intake increased the production of inflammatory markers and reduced the richness and diversity of the gut microbiota in *H. pylori*-infected mice. Further analysis showed that TMAO enhanced the relative abundance of*Escherichia_Shigella* in *H. pylori*-infected mice, which had positive correlations with the levels of LPS, IL-6, and CXCL1. Collectively, our results suggest that TMAO may promote *H. pylori*-induced inflammation by increasing the viability and virulence of *H. pylori* and may aggravate inflammation in association with the gut microbiota in *H. pylori*-infected mice. This study may provide a novel insight into the mechanism for the effect of diet-derived metabolites such as TMAO on *H. pylori*-induced disease development.

Enantiomeric Glycosylated Cationic Block Co-Beta-Peptides Eradicate Antibiotic-Tolerant Persisters and Biofilm in Staphylococcus Aureus

FOODIE Asia

FOOD INNOVATION AND ENGINEERING CONFERENCE

Conference: Food Innovation and Engineering Asia Conference

Year: 2019

Proceeding: 2019 Food Innovation and Engineering (FOODIE) Asia Conference

Group: Poster Session

Session: Poster Session

Authors:

Zhang, K., NTU

The treatment of bacterial infection is hindered by the presence of biofilm and metabolically inactive persisters. Methicillin-resistant Staphylococcus aureus (MRSA) is a major pathogen causing mortality and morbidity due to infections. We have synthesized a new enantiomeric block co-beta-peptide, (PDGu(7)-b-PBLK(13)), with high yield and purity by a one-shot one-pot anionic-ring opening (co)polymerization (AROP). The co-beta-peptide is bactericidal against replicating as well as biofilm and persisters MRSA, and also disperses biofilm biomass. Its antibacterial activity is superior to vancomycin in established MRSA murine and human ex vivo skin infection models, with no acute in vivo toxicity in repeated dosing in mice at above therapeutic levels. This new class of non-toxic molecule, effective against all bacterial sub-populations, has promising clinical potential.

Glucose-Derived Cationic Block Poly(?-peptides) for Potentiating Antibiotics Against Gram-Negative Bacteria

Conference: Food Innovation and Engineering Asia Conference

Year: 2019

Proceeding: 2019 Food Innovation and Engineering (FOODIE) Asia Conference

Group: Poster Session

Session: Poster Session

Authors:

Si, Z., Nanyang Technological University

Carbapenem-resistant gram-negative bacteria are at the top of The World Health Organization (WHO) list of pathogens for which antibiotics are critically needed. In addition of their propensity to acquire antimicrobial resistance genes through horizontal gene transfer, the presence of an outer membrane and multiple efflux pump systems confers a high degree of intrinsic drug resistant in gram-negative pathogens. Many antibiotics are either unable to penetrate the outer membrane layer, or are excluded from the cytosol by efflux mechanisms.

Integration of Sustainability Assessment and Quantitative Risk Assessment of Aquaculture in Singapore

FOODIE Asia
FOOD INNOVATION AND ENGINEERING CONFERENCE

Conference: Food Innovation and Engineering Asia Conference

Year: 2019

Proceeding: 2019 Food Innovation and Engineering (FOODIE) Asia Conference

Group: Poster Session

Session: Poster Session

Authors:

Glendon, O. H. M., Nanyang Technological University Food Technology Centre (NAFTEC), School of Chemical and Biomedical Engineering, Nanyang technological University, Singapore

This project is a partnership between Nanyang Technological University, NAFTEC Food Technology Centre, and the Technical University of Denmark (DTU), division for Quantitative Sustainability Assessment, and is part of a bigger project about antimicrobial resistance and sustainability assessment of the aquaculture industry. The principal goal of this project is to assess aquaculture impacts on human health and the environment. The world needs a safe and sustainable food system that can feed every person every day with enough quantity and good quality. Being a small nation with limited land resource and large seafood consumption, Singapore needs to emphasize the assessment of the safety and sustainability of our aquaculture production. The interdisciplinary nature of the project focuses on food systems for safe and sustainable aquaculture development, and the development of science-based assessments as a basis for decision support for regulatory agencies and relevant industries. The aquaculture industry is associated to several environmental impacts, such as biodiversity loss, climate change and eutrophication of water bodies. Furthermore, the misuse of antibiotics for seafood production for growth promotion has resulted in an increase in antimicrobial resistance in the microbiome of the seafood, which can be an important threat for human health according to the World Health Organization. Therefore, it is important that the development of aquaculture happens in the most sustainable and safest way possible. A common tool to assess food safety is quantitative microbial risk assessment while the common tool to assess aquaculture sustainability is life cycle assessment. Both assessments when combined, are scientific decision support tools that can be used as a significant leverage to influence high-level policy making in the aquaculture industry to ensure environmental sustainability and food safety.

Lactic Acid Bacteria As Microbial Cell Factories: Updating the Heterologous Expression Toolbox

FOODIE Asia

FOOD INNOVATION AND ENGINEERING CONFERENCE

Conference: Food Innovation and Engineering Asia Conference

Year: 2019

Proceeding: 2019 Food Innovation and Engineering (FOODIE) Asia Conference

Group: Poster Session

Session: Poster Session

Authors:

Wong, F. T., Molecular Engineering Laboratory, A*STAR
Ow, D. S. W., Bioprocessing Technology Institute, A*STAR
Tan, L. L., Molecular Engineering Laboratory, A*STAR
Lim, P., Bioprocessing Technology Institute

Lactic acid bacteria (LAB) are a "generally regarded as safe" microorganisms which are traditionally used in food fermentation and as chassis for heterologous protein production. The order of LAB include mainly *Lactobacillus*, *Leuconostoc*, *Pediococcus*, *Lactococcus*, and*Streptococcus*. In recent years, they have also emerged as gut delivery tools and cell factories for protein therapeutics and within food fermentation. However, these applications are heavily depended on the secretion efficiencies and capabilities of LAB. To facilitate rapid and productive heterologous expression, an efficient recombinant protein expression toolbox is required.

In our labs, an entire suite of tools â€“ ranging from promoters, chaperones, tags and solubility partners - have been examined to accelerate the design, build and test process of optimizing recombinant protein expression. In this presentation, we will focus on (1) the comparison of secretion tags and new solubility partners and (2) also describe new propeptide sequences for enhancing solubility and secretion efficiencies in LAB.

Microencapsulation of PUFA Enriched Ulva Lactucaâ Oil for Commercial Applications

Conference: Food Innovation and Engineering Asia Conference

Year: 2019

Proceeding: 2019 Food Innovation and Engineering (FOODIE) Asia Conference

Group: Poster Session

Session: Poster Session

Authors:

I, A. E. S. Jr., Vel Tech High Tech Dr Rangarajan Dr Sakunthala Engineering College,Affilated to Anna
Jayabalan, V. Sr., Vel Tech High Tech Dr Rangarajan Dr Sakunthala Engineering College,Affilated to Anna

Essential oils are the volatile lipophilic compounds which are unable to synthesize by mammals, therefore should be obtained through diet. The objective of this study is to microencapsulate the PUFA enriched algal oil extracted from *Ulva lactuca*. The proximate analysis of algal powder shows that it was characterized with a high content of carbohydrates (55 %), protein (22 %), fat (1.7%), antioxidant activity (34.6%) and moisture content (0.42%). Different lipid extraction strategies have been followed and the highest lipid extraction of 28.13% v/v was achieved by Soxhlet extraction using n-hexane. Further, various antimicrobial studies as well as oxidative stability of the extracted oil was also evaluated. For commercial applications, the optimized microsphere with 6% alginate concentration, 4% cross linking agent and 15 min of the cross linking time were observed to have a sustained in vitro release profile. At pH 6.8, approximately 90 % of the encapsulated algal oil was released for a period of 48 hours. Additional coating of the encapsulated oil without antioxidant shows a shelf life of 24 days, with 0.1 % tocopherol shows maximum 32 days and 0.05% chitosan coated capsule shows the maximum 36 days of shelf life time.

Key words: Essential oil; *Ulva lactuca*; Anti-oxidants; Microencapsulation; Shelf life

Optimization of Various Edible Oil By Determining Totox Value Using Natural Antioxidants

Conference: Food Innovation and Engineering Asia Conference

Year: 2019

Proceeding: 2019 Food Innovation and Engineering (FOODIE) Asia Conference

Group: Poster Session

Session: Poster Session

Authors:

Jayabalan, V. Sr., Vel Tech High Tech Dr Rangarajan Dr Sakunthala Engineering College,Affilated to Anna
I, A. E. S. Jr., Vel Tech High Tech Dr Rangarajan Dr Sakunthala Engineering College,Affilated to Anna

The research work has been aimed to compare the effects of various natural antioxidants with two different edible oils such as palm oil, Refined sunflower oil and Rice bran oil. Recent literature survey reveals that the synthetic antioxidants have been confirmed for their toxicological and carcinogenic effects. Therefore addition of natural antioxidants shows significant effects as compared with synthetic ones and hence this has paved the way for the identification of various sources of natural antioxidants such as ginger, grapes seed, cinnamon and cloves. The selected natural antioxidants have been extracted through the Soxhlet apparatus method and their kinetic studies are also determined. The potential implications of all the natural antioxidants with high oxidative stability have been analyzed and evaluated by various values such as peroxides values, p-Anisidine values and Totox value etc., in both the edible oils through Response Surface Methodology(RSM). The great challenge for this field has been discussed briefly in all aspects along with current studies and it has been provided in-depth information related to health effects and its remedial causes.

Keywords

Natural Antioxidants, Edible Oil, Oxidative Stability, Rancimat Analyzer, Response Surface Methodology (RSM).

Phenotypic and Genotypic Characterization of Antimicrobial Resistant Escherichia coli isolated from Ready-to-Eat Food in Singapore Using Disk Diffusion, Broth Microdilution and Whole Genome Sequencing Methods

FOOD INNOVATION AND ENGINEERING CONFERENCE

Conference: Food Innovation and Engineering Asia Conference

Year: 2019

Proceeding: 2019 Food Innovation and Engineering (FOODIE) Asia Conference

Group: Poster Session

Session: Poster Session

Authors:

Guo, S., Nanyang Technological University
Feng, M. T. Y., NTU Food Technology Centre
Aung, K. T., Nanyang Technological University
Schlundt, J., Nanyang Technological University

Antimicrobial resistance (AMR), especially multidrug-resistance, of bacteria is posing a great threat to public health. This study aimed to determine the antimicrobial resistance profiles of *Escherichia coli*isolated from ready-to-eat food sold in retail food premises in Singapore. In this study, a total of 99 *E. coli* isolates from poultry-based dishes (n=77) and fish-based dishes (n=22), obtained between 2009 and 2014, were included for disk diffusion testing. Of the 99 isolates, 24 (24.2%) were resistant to at least one antimicrobial agent. These isolates were then subjected to broth microdilution testing against 33 antimicrobial agents to determine the minimum inhibitory concentration (MIC) of isolates. Finally whole genome sequence (WGS) was carried out on the strains in order to correlate resistant phenotypes to putative antimicrobial-related genes. Of the 24 isolates, 15 (62.5%) were found to be resistant to three or more classes of antimicrobials and thus were defined as multi-drug resistant strains. Two isolates (8.3%) were confirmed as Extended-Spectrum Î²-lactamase (ESBL)- producing *E. coli*by double disk synergy test. By analyzing the WGS contigs using BLASTn and KmerFinder, ESBL genes and transferable colistin resistance gene*mcr*-1 and *mcr*-5 were determined to be located on plasmids, which could pose a greater risk of AMR transfer among bacteria. This study showed the presence of antimicrobial resistant *E. coli* isolates in ready-to-eat retail food, and raises a concern on the possible transmission of antimicrobial resistant bacteria from food to humans.

AUTHOR INDEX